Alachua County ARES

Waccasassa Wildfire Exercise

(June 9, 2018)

After Action Report/
Improvement Plan

June 10, 2018

by John Troupe KM4JTE

Edited by Gordon L. Gibby KX4Z

ISBN-13: 978-1721727810

ISBN-10: 1721727817

Exercise
Location and
satellite photo

CONTENTS

ACKNOWLEDGMENTS

The Alachua County ARES group would like to acknowledge all who helped make this activity possible, particularly our own volunteers who worked so hard for so many months, and also the Alachua County EOC which has supported our efforts at backup emergency communications by purchasing multiple radios and large storage batteries. We particularly acknowledge:

Mr. Mike Sherrill of Florida Forest Service , who provided the site and the tower for amateur use;

Lt. Kevin Rulapaugh, MARC Unit 3, Alachua Fire Rescue, who demonstrated the MARC Unit capabilities and potential involvement of amateur radio.

Comm Station 1 and Incident Command
Massive towers participating courtesy of Florida Forest Service and Region 3 MARC Unit.
(The little white travel trailer functioned for Incident Command.)

FOREWORD

Our Alachua County Amateur Radio Emergency Service group has persevered some three years now developing radically new *skills, assets and strategies*. In order to meet our needs, we created two new amateur radio clubs – the North Florida Amateur Radio Club (call: NF4RC) and the Alachua County EOC Radio Club (call: NF4AC) to serve our communications needs, including WINLINK authorized callsigns. Our primary educational web page is http://www.qsl.net/nf4rc/ but we also maintain http://www.qsl.net/nf4ac/. Most of our members also serve in the Gainesville Amateur Radio Society, a club with a broader mission than our emergency-oriented organization.

Roughly two years ago we began to hold what we call "Full Scale Exercises" involving real radios, deployment to diverse locations, and structured practice of simulated emergency situations. After a year of getting used to these, we've now begun a process of "broadening our leadership." It isn't healthy for one or two people to always be the ones to plan and carry out leadership of growing, vibrant emergency-oriented amateur radio clubs! So we've begun the process of having "new blood" take the reins of our group – and John Troupe KM4JTE, with the able co-leadership of Vann Chesney AC4QS, took on the challenge of managing our group's "Spring Exercise."

John was especially interested in carrying out a "wildfire" simulation, and with his strong ham background in HF communications, and his recent fascination with narrowband digital (PSK31) communications, he wrote a scenario designed to give the group some fairly straightforward messages to send from deployed positions.

In the process, a huge amount of growth occurred! Jeff Capehart W4UFL has always indicated that *just getting ready for an Exercise* is where a lot of the benefit occurs. At our practice "tabletop" (with full radios) there were lots of problems with radio configuration, RFI and other issues. In order to create the ICS documentation for an exercise, John (an Apple computer guy) had to learn how to transfer files to more pedestrian WINDOWS users. All of us gained expertise with radios, antennas, and the intersection with modern office computing abilities. In the midst of this growth, our group did a LunchNLab group-build of eleven off-center-fed antenna baluns --- contributing more assets that could be used in this exercise. To add the final layer, John pushed to build and strengthen ties with the Florida Forest Service and also with the State of Florida Region 3 MARC unit. This all strengthened the interoperability capabilities of our Alachua County ARES group.

The Exercise went well --- as you'll read in these pages --- and then John did a wonderful job putting together this After Action Report. A job well done, and two leaders stronger and better trained as a result!

Gordon L.Gibby MD KX4Z
Newberry, Florida
June 2018

SECTION 1: EXERCISE OVERVIEW

Exercise Details

Exercise Name
2018 Waccasassa Wildfire Exercise (Simulated Emergency Test)

Type of Exercise
Full Scale Exercise

Exercise Start Date
June 9, 2018

Exercise End Date
June 9, 2018

Duration
5 Hours

Location
Waccasassa Forestry Center, Gainesville, Florida

Sponsor
Alachua County ARES, a component of the American Radio Relay League (ARRL)

Program
Amateur Radio Emergency Service

Mission
Communications Support

Capabilities
VHF local communications, analog voice

HF local and national communications, digital WINLINK.

HF local PSK-31

ICS Forms

Scenario Type

Communications Failures

Exercise Planning Team

John Troupe, KM4JTE

Vann Chesney, AC4QS

Participating Organizations

Alachua County, Florida (Emergency Operations Center)

Alachua County ARES

Florida Forest Service

MARC Unit 3

Alachua Fire Rescue

Number of Participants
- Players - 12
- Controllers - 0
- Evaluators – 0

Morning Briefing   just before start of exercise.

SECTION 2: EXERCISE DESIGN SUMMARY

Exercise Purpose and Design

Natural disasters, including wildfires, produce extreme burdens on communications, especially when power is out, when cell towers are down and when emergency services need to transfer excessively large numbers of messages between agencies. The recent Thomas Fire in Southern California, which burned 281,000 acres (440 square miles), demonstrated the scale of potential damage from wildfires. (See: https://en.wikipedia.org/wiki/Thomas_Fire) Florida, with its lush vegetation, spring droughts, and large areas of wildland-urban interface from increasing population, is also at risk for catastrophic wildfires. In 1998, fires ravaged the state and forced the evacuation of hundreds of thousands of residents.

Our volunteers have recognized the value of WINLINK email for inter-agency communications, and one of our volunteers' stations was used to assist as a WINLINK email relay station for Puerto Rico. The group has used both VHF packet and HF to access the WINLINK network. We have also trained with keyboard-to-keyboard communications via digital modes, specifically, PSK-31.

Our volunteers have recognized the need to practice setting up stations without commercial power, providing digital messaging between isolated stations and the Emergency Operations center and other agencies. The importance of using ICS forms, including the ICS-214, was recognized and participants documented using the 2018 Alachua County Amateur Radio Emergency Communications Reference Forms & Reference Material compiled and published by Gordon Gibby KX4Z. (See: https://www.amazon.com/Alachua-Amateur-Emergency-Communications-Reference/dp/1718678029/)

This exercise followed Incident Command System procedures, as in previous exercises.   In this exercise we wanted to further our training by

- deploying to a location in cooperation with other agencies, and using mobile antennas systems without external power
- setting up with portable antenna systems and battery power
- providing communications using repeater and relayed VHF
- testing HF long-distance communications using WINLINK and PSK-31

Training for this type of scenario – total loss or overwhelming of local conventional communications systems (telephone/Internet) has been an ongoing program of our club for over three years.

Development of this exercise began months before, with ongoing training, and meeting with cooperative agencies.

An Exercise Plan was created with details for participants, over three months before the event.   All relevant ICS forms were created.

Exercise Objectives

- **Objective 1: Assess the capabilities of our group to set up both mobile and portable HF and VHF antenna systems in an environment without power or cell service**

  Capability: ANTENNA PLACEMENT, MOBILE DEPLOYMENT, Portable Deployment

- **Objective 2: Provide practice for, and assess our capabilities at sending/receiving WINLINK email messages and attachments HF and/or VHF**

  Capability: WINLINK COMMUNICATIONS, BACKUP POWER, MOBILE DEPLOYMENT, PORTABLE

- **Objective 3: Assess our capabilities at sending and receiving digital keyboard to keyboard messages and bulletins on HF (e.g. PSK-31)**

  Capability: KEYBOARD TO KEYBOARD DIGITAL COMMUNICATIONS

- **Objective 4: Assess our capabilities at utilizing the ICS system, including the processing of ICS-214 and other relevant ICS forms**

  Capability: Skills in using ICS forms and the ICS system

Scenario Summary

June 6, 2018

"The past fall had record rainfall and conditions were perfect for abundant growth of vegetation in

the region. Farmers, foresters and nature lovers were thrilled with the green. Unfortunately, this was followed by record drought throughout the winter and spring. This was a trend throughout North America, and wildfire fighting resources are busy in several active fires. The Florida Department of Agriculture and Consumer Resources has been distributing " Wildfire Reduction in Florida" to community leaders and has encouraged its reading on local broadcasts. In Alachua County, the Keetch-Byron Index is at a record low of 700 and the fire index is Extreme. In addition to drought, lightning has been extreme, and arson is the suspected cause of several fires. There are several fires in neighboring Putnam County, including a smoldering fire in Levy Prairie. Other areas of significant concern are the wildfires near LaCross and in the Morningside Park in Alachua County. With spreading of the fires, operability of repeaters and broadcast stations in northwest Gainesville, is threatened. The Deerfield Power Plant is also in the path. Smoke- induced respiratory conditions and poor visibility are already problems. Anticipating an overwhelmed cellphone service, as well as emergency communications, and with likely evacuations, Alachua County ARES has been asked to assist with communications, especially between the Florida Forest Service and the Emergency Operations Center. Alachua Fire Rescue has deployed it's MARC Unit 3. "Last mile" communications to and from evacuation evacuation staging areas will also be needed. While our presence at the fire line is NOT requested, we are asked to set up operations at the Waccasassa Forest Center, where we will be in proximity to Forest Service communications. The evacuation staging area is simulated within the compound. Forestry will coordinate firefighting operations, and the EOC/County Sheriff will coordinate evacuation from threatened areas. "

We plan to establish two communications stations under Incident Command by Gordon Gibby KX4Z. Comm Station 1 will primarily send and receive traffic between the Forestry and the EOC with operators Jeff Capehart W4UFL and Susan Halbert KG4VWI. Comm Station 2 will coordinate communications between the EOC and the evacuation staging area. 2m HT's will be used between the comm stations and, when possible, for tactical messages between the comm stations and the EOC. In the event of inoperable VHF repeaters, a relay, designated Comm Station 3, will be assigned.

At 0900, exercise participants assembled at the Waccasassa Forestry Center and received a briefing including sign-in on ICS-211, a review of the ICS-201, introduction of participants, setting of simplex frequency to HT's, safety and housekeeping instructions, and dividing into teams. Comm Stations received envelopes providing status reports and related tasks. The first Envelopes were opened at 1000 and the second at about 1100.

Envelope 1:

Comm Station 1:

Forestry Status Report:

Out of control wildfires north of Gainesville are rapidly heading to the south. In the next 24 hours the fire is expected to reach the areas north of NW 39th AVE (HWY 222) and west of NW 13th ST (HWY 441). I 75, HWY 441 and other roads to the north of the this area have been closed due to fire and smoke. Evacuation routes need to be arranged to the south and west.

Comm. 1 Task:

- Inform the EOC of these conditions via Winlink form ICS-213 using frequencies outlined in the ICS-205 form. Verify reception via the EOC voice channel.

Note: The above status report will be given verbally as if from an informal briefing. The message sent will be a summarized version of the report.

Comm Station 2:

Staging Area Status Report:

A Voluntary evacuation order has been issued for NW Gainesville and the surrounding areas. An evacuation staging area has been established at the Forestry Center for people without transportation.

Currently 40 people have reported to and requested transportation from this location. All are ambulatory and without medical complaints. More are anticipated, we will need temporary shelter, drinking water and MRE's for 100 people. Arrangements need to be made for transporting all evacuees at this location to shelters outside the affected area.

Comm 2 Task:

- Inform the EOC of these conditions via Winlink form ICS-213 using frequencies outlined in the ICS-205 form. Verify reception via the EOC voice channel.

Envelope 2:

Comm Station 1:

Forestry Status Report:

The fires to the north of Gainesville have progressed more rapidly than expected. There is widespread damage to homes and businesses along Millhopper RD and at NW 39th AVE and 143rd ST. The fire continues moving south. The fire continues moving south. Due to the fires in NW Gainesville, power, utilities, phone, internet and cell services are down. All amateur radio repeaters are down.

Task 1:

- Establish a voice communications link to the EOC via a simplex relay station to be named Comm station 3. Direct all voice traffic to the EOC via Comm. 3.

Forestry Status Report:

A fire (thought to be arson) is developing northeast of Gainesville, west of Hwy 301 near Hwy 26. HW 26 west of HW 301 has been closed for westbound traffic. The fire has taken out the closest Winlink RMS station KX4Z. Due to poor band conditions, it is impossible to connect to any Winlink station.

Task 2:

- Coordinate with the EOC via Comm. 3 to setup a digital communications link using PSK-31. Send the following information to the EOC:

  A fire (thought to be arson) is developing northeast of Gainesville, west of Hwy 301 near Hwy 26. HWY 26 west of HWY 301 has been closed for westbound traffic.

Envelope 2:

Comm Station 2:

Forestry Status Report:

The fires to the north of Gainesville have progressed more rapidly than expected. There is widespread damage to homes and businesses along Millhopper RD and at NW 39th AVE and 143rd ST. The fire continues moving south. Due to the fires in NW Gainesville power, utilities, phone, internet and cell services are down. All amateur radio repeaters are down.

Task 1:

The fire has taken out the closest Winlink RMS station KX4Z. Due to poor band conditions, it is impossible to connect to any Winlink station.

- Establish a voice communications link to the EOC via a simplex relay station to be named

17

Comm. 3. Direct all voice traffic to the EOC via Comm. 3.

Task 2:

Forestry Status Report:

A fire (thought to be arson) is developing northeast of Gainesville, west of Hwy 301 near Hwy 26. HWY 26 west of HWY 301 has been closed for westbound traffic.

- Coordinate with the EOC via Comm 3 to setup a digital communications link using PSK-31. Send the following information to the EOC:

    100 evacuees in staging area, 5 are in respiratory distress

    Send ambulances and medical personnel to transport 5 patients

    Send 100 gallons of drinking water and 100 MREs

    Send buses to evacuate 100 people

(L to R) Alvin Osmena KM4DLF, Mike Shaffer KD4INH, Leland Gallup AA3YB and John Troupe KM4JTE at Comm Unit 2 using end-fed long wires.

SECTION 3: ANALYSIS OF CAPABILITIES

This section of the report reviews the performance of the exercised capabilities, In this section, observations are organized by capability and associated activities. The capabilities linked to the exercise objectives Waccasassa Wildfire Exercise are listed below, followed by corresponding activities. Each activity is followed by related observations, which include analysis, and recommendations.

CAPABILITY 1: ANTENNA PLACEMENT, MOBILE DEPLOYMENT, PORTABLE DEPLOYMENT

CAPABILITY SUMMARY: Fixed, pre-existing antennas should obviously perform for necessary communications. However, when these are not present, mobile towers provided by agencies such as the Florida Forest Service or Alachua Fire Rescue MARC are often even better. Towers that can be lifted to 100 ft while guyed and 60 ft unguyed provide support for VHF and wire HF antennas. When no towers are available, volunteers need to have the skills to launch wire antennas into trees or other support.

OBSERVATIONS AND ANALYSIS:

Very close proximity of the incident commander HF station meant that coordination of frequencies used by powerful transmitters had to be arranged over handitalkies to avoid damage to receivers. **Antennas placed further apart is preferable.**

Comm station 2, the evacuation staging area, was not easily accessible by autos and had no generator or tower. Operations were conducted in a three- sided shed under an oak tree, which supported end-fed antennas, without excellent "ground" return path. Radio Frequency Interference locked up the computer's USB-port driver, leaving the transmitter in transmit mode.

Solutions include:

1. Use the "most balanced" antenna systems possible.   Center fed dipoles are easiest.
2. Keep radiating antennas away from computers as much as possible.
3. Use current-baluns between antenna and radio equipment to try and avoid "ground" currents following as common mode currents on coax or balanced line.
4. Use ferrite beads on the audio cables from the radio to the soundcard interface.

5. Use ferrite beads on the USB cable from the soundcard interface and the computer.
6. Consider a ferrite bead on the USB cable from the external mouse.
7. Use TWO current baluns in the transmission line to the transmitter.
8. Reduce power if the above doesn't work.

Recommendations:
Continue practice in antenna installation in the field, with and without towers, serving digital communications.

CAPABILITY 2: WINLINK COMMUNICATIONS, BACKUP POWER, PORTABLE DEPLOYMENT

A. Capability Summary: WINLINK provides a world-wide, radio-based email capability that has been leveraged by mariners, emergency communications personnel, missionaries, and the Federal Government. Allowing both email and attachments, it can speed digital messages toward areas where the Internet is still working, and then forward them by the far-faster internet email facilities, or in a complete national disaster, can slowly move email to "Message Pickup Stations" by radio alone. It is the premier HF radio-based email system in the world today.

OBSERVATIONS AND ANALYSIS:

The Comm stations had difficulty in reaching HF WINLINK Gateways. [Ed. Note: There were ZERO sunspots on the Exercise day, and for at least 4 prior days, with SFI = 67 on Exercise day---truly difficult situations for HF communications.] There were some difficulties with transmitters in close proximity, and WINLINK connections tend to be the HARDEST right in the middle of the day. The number of stations in "green" propagation prediction dwindles badly. HF contact with WINLINK gateways tends to be EASY in the morning, evening, and dark hours but on lower bands (80/40/30) while during the day 40/30/20/18 may have to be used.

When unable to reach distant HF gateways, the Incident Commander employed the local VHF packet system to send/receive WINLINK via KX4Z-11. (145.030 to Beatty Towers W4DFU-8...digipeating directly to KX4Z-11 on 145.030). Some difficulty was initially noted with KX4Z-10.

Improvement involves:
a) detailed geographic knowledge of 2 meter packet systems is a requirement to best use them
b) possibly still some internet connectivity issues at the KX4Z station (AT&T DSL) – switching to KX4Z-11 (which caches to RMS-RELAY on a windows machine instead of direct to the internet) got around this issue and email flowed fine
c) better knowledge of packet systems allows knowledge of when digipeater operations will work and when a CONNECT script will be needed to switch ports to change frequencies.

RADIO-ONLY Tests

Dave Welker W2SRP graciously sent us radio-only WINLINK emails. The transit of Dave Welker's email was interesting: he had trouble even getting it off to a radio-only gateway due to his own antenna/propagation issues, but finally got it to a Canadian gateway, VE1YZ, which immediately attempted a direct connection to one of common MPS stations, KX4Z gateway --- this made an initial connection, then the propagation failed so the Canadian gateway computed a relay route and started the message on it's way. It reached Gainesville in 1 hour 9 minutes, after multiple automated relays.

Recommendations:

- Continue WINLINK practice. An opportunity is the Florida WINLINK NET.
- Continue practice on Peer-to-peer WINLINK skills
- Continue to train on radio-only WINLINK emails

BACKUP POWER AND PORTABLE DEPLOYMENT

B. Capability Summary:

Electrical Utility power loss is one of the most frequent occurrences in hurricanes, and is a major cause of loss of traditional communications.   Amateur radio emergency volunteers need to have alternate power capabilities.

OBSERVATIONS AND ANALYSIS

ALL stations active in this Exercise operated out of battery or generator power throughout the entire exercise. Comm station 1 and Incident command worked off generators. Comm station 2 worked off batteries, one of which was a deep-cycle battery connected to power cables with wrong size eyelet holes. Power systems should be planned in advance to avoid unexpected mismatches. Polarity protection should be placed permanently on wires to expensive transceivers.

Recommendations
- Practice mobile deployment, with attention to grounding.

VHF Relay

While not a cited core capability, our group simulated an inoperable repeater and the need to relay voice

23

messages to the EOC. This was capably done by Stuart Kaye WZ0T, tactical call Comm Station 3.

CAPABILITY 3: KEYBOARD TO KEYBOARD DIGITAL COMMUNICATIONS

Capability Summary: Keyboard to keyboard digital modes, such as PSK-31 allows for communication with low power and with a bandwidth even narrower than with CW. This mode can also send ICS forms and bulletins. Unlike WINLINK peer to peer, with places the messages in an inbox of the recipient, messages with PSK-31 lack this confidentiality.

OBSERVATIONS AND ANALYSIS:

PSK-31 messages were successfully sent and received on the 10 meter band. The EOC, even with a weak antenna, received, and confirmed. The parties were about 3 miles apart, and it was likely that our transmissions were direct (point to point) or ground plane. However, we did observe QSO's from stations in Spain and Sierra Leone on 20 meters.   This is fortunate, as HF propagation was poor for sending Winlink Winmor messages.

Unfamiliarity with FLDIGI – led to operators entering an entire message, before initiating transmission, leaving "dead air" on the frequency so the counterparty thought the link was lost.

Unless battery power is an issue, keep a signal up as fast as possible --- hit the TX or T/R button immediately when transmitting and begin typing on PSK31 --- doesn't matter if you get ahead or behind, or even use backspace --- it works perfectly.

Recommendations:

- Continue practice with PSK-31
- Investigate FLMSG, which enables transmission of prepared ICS forms.

CAPABILITY 4: ICS FORMS

Capability Summary: ARES volunteers have been becoming more accustomed to standard FEMA/ICS forms at previous simulation events. Importance of the forms for planning, documentation of tasks, and improvement are recognized by the group.

OBSERVATION AND ANALYSIS

Planning documents including ICS 201, 202, 203, 204, 205, 206 and 208 were started months in advance. ICS forms 211, 213, 214 and 309 were all completed during the exercise. Knowledge of and access to the forms was facilitated by Alachua ARES Forms & Reference Material by Gordon Gibby, KX4Z, distributed before and at the exercise. Some of the ICS 214 forms, activity logs, were duplicative of the 309, communications log.

Recommendation:

- Training specific to the ICS-214 planned per 2018 EOC Training and Hurricane Response Tabletop Exercise, date to be announced

SECTION 4: CONCLUSION

The Waccasassa Wildfire Exercise was conducted on June 9, 2018, to test Alachua County ARES capabilities to provide backup emergency communications to a simulated, area wildfire.

While not the most ambitious Full Scale Exercise carried out by Alachua County ARES, it served as an opportunity to hone our skills in establishing and operating emergency stations where we used digital modes we had used, and one we had not used in a full -scale exercise. We also simulated a VHF repeater-down situation to which we responded with a relay station.   A very wide array of communications skills were put to the test, including simplex VHF voice, simplex VHF voice, relayed VHF, HF WINLINK, and PSK-31.

These skills (WINLINK, HF) are not universally practiced within our group, and this exercise solidified their usage of them.   Furthermore, we strenuously tested their abilities to place emergency antennas and provide alternative power -things that would be important in a real wildfire / communications emergency. Our group performed admirably at these tasks.   In addition, our group conscientiously utilized the ICS forms. Many performance issues were external to our group- the HF propagation was abysmal, but our group got the messages through. These were significant advancements for our group. This was the very first time that our group ever used keyboard-to-keyboard HF modes (here, PSK-31), during an exercise.

This exercise allowed us to begin to put together our own internal Incident Command System, and the entire group learned about the planning and execution of exercises, which helped broaden our leadership involvement.

Weaknesses of our deployment included demonstrations of improperly grounded antennas, poor power supply connections, non-standard software without a backup, and some clumsy exchanges. These are deficits which will be corrected with continued training, practice using the equipment with which we will deploy, and more exercises.

Critical to the success of this exercise was the support we had from Mike Sherrill of Florida Forest Service and Lt Kevin Rulapaugh MARC unit 3, Alachua Fire Rescue. Many thanks for the use of the site, use of antenna towers, and, from Kevin, the demonstration MARC unit, the capabilities we hope to help.

APPENDIX A: ISSUES NOTED / IMPROVEMENT PLAN

(updates ongoing at: http://qsl.net/nf4rc)

Updated as of June 11 2017

No.	Issue	Suggestion	Actual Action Taken
1	End-fed antennas, without excellent "ground" return path. Radio Frequency Interference locked up usb driver, leaving the transmitter in transmit mode.	1. Use the "most balanced" antenna systems possible. 2. Keep radiating antennas away 3. Use current-baluns between antenna and radio equipment 4. Use ferrite beads on the audio cables from the radio to the soundcard interface. 5. Use ferrite beads on the USB cable from the soundcard interface and the computer. 6. Use TWO current baluns in the transmission line to the transmitter. 7. Reduce power	Immediate problem RESOLVED Power reduced Longer Term Solutions: 1. Use appropriate grounding on antennas in subsequent exercises and use balanced antennas when possible 2. Recognize that digital modes need low power, typically 20 watts or less

2	Deep cycle battery connected to power cables with wrong size eyelet holes.	1. Power systems should be planned in advance to avoid unexpected mismatches. 2. Polarity protection should be placed permanently on wires to expensive transceivers.	Long term solution is workshop involving power connections, and further exercises using battery power
3	While using FLDIGI, operators entered an entire messages, leaving "dead air" on the frequency so the counterparty thought the link was lost.	Unless battery power is an issue, keep a signal up as fast as possible --- hit the TX or T/R button immediately when transmitting and begin typing on PSK31 --- doesn't matter if you get ahead or behind, or even use backspace.	Solutions 1. Continue practice with PSK-31 in future exercises 2. Investigate FLMSG, which enables transmission of prepared ICS forms.
4	Some of the ICS 214 forms, activity logs, were duplicative of the 309, communications log.	The Activity log (ICS-214) records details of notable activities at any ICS level. These logs provide basic incident activity documentation, and a reference for any After Action Reports.	Training specific to the ICS-214 planned per 2018 EOC Training and Hurricane Response Tabletop Exercise, date to be announced

APPENDIX B: LESSONS LEARNED

While the After Action Report/Improvement Plan includes recommendations which support development of specific post-exercise corrective actions, exercises may also reveal lessons learned which can be shared with the broader homeland security audience. Federal Emergency Management Agency (FEMA) maintains the *Lessons Learned Information Sharing* (LLIS.gov) system as a means of sharing post-exercise lessons learned with the emergency response community. This appendix provides jurisdictions and organizations with an opportunity to nominate lessons learned from exercises for sharing on *LLIS.gov*.

For reference, the following are the categories and definitions used in LLIS.gov:

- **Lesson Learned:** Knowledge and experience, positive or negative, derived from actual incidents, such as the 9/11 attacks and Hurricane Katrina, as well as those derived from observations and historical study of operations, training, and exercises.

- **Best Practices:** Exemplary, peer-validated techniques, procedures, good ideas, or solutions that work and are solidly grounded in actual operations, training, and exercise experience.

- **Good Stories:** Exemplary, but non-peer-validated, initiatives (implemented by various jurisdictions) that have shown success in their specific environments and that may provide useful information to other communities and organizations.

- **Practice Note:** A brief description of innovative practices, procedures, methods, programs, or tactics that an organization uses to adapt to changing conditions or to overcome an obstacle or challenge.

Exercise Lessons Learned

The following subject headings are lessons derived from the Alachua County, Florida FSE on May 6, 2017 that are proposed for inclusion in the Department of Homeland Security's Lessons Learned/Best Practices web portal, LLIS.gov:

- THE ASSISTANCE OF STATE, LOCAL, AND PRIVATE ENTITIES CONTRIBUTED GREATLY TO THE LEARNING OPPORTUNITIES AFFORDED BY OUR EXERCISE.
- IN ADDITION TO INCORPORATING NOVEL MODES AND TECHNIQUES INTO OUR CAPABILITIES, ONGOING TRAINING AND CONTINUED EXERCISES ARE ESSENTIAL TO OUR ABILITY TO RESPOND

Two 100-foot towers and a refurbished travel trailer with a fiberglass mast. All with air conditioning.

www.ingramcontent.com/pod-product-compliance
Lightning Source LLC
Chambersburg PA
CBHW081650220526
45468CB00009B/2604